J. Henry Gilbert

Introduction to the Study of the Scientific Principles of Agriculture

being the inaugural lecture, delivered May 6, 1884, at the University museum, Oxford

J. Henry Gilbert

Introduction to the Study of the Scientific Principles of Agriculture
being the inaugural lecture, delivered May 6, 1884, at the University museum, Oxford

ISBN/EAN: 9783337313418

Printed in Europe, USA, Canada, Australia, Japan

Cover: Foto ©berggeist007 / pixelio.de

More available books at **www.hansebooks.com**

INTRODUCTION

TO THE STUDY OF

THE SCIENTIFIC PRINCIPLES

OF

AGRICULTURE;

BEING

The Inaugural Lecture,

DELIVERED MAY 6, 1884,

AT THE UNIVERSITY MUSEUM, OXFORD,

BY

JOSEPH HENRY GILBERT, M.A., Ph.D., LL.D.,

**SIBTHORPIAN PROFESSOR OF RURAL ECONOMY IN THE
UNIVERSITY OF OXFORD;**

*Vice-President of the Chemical Society; Fellow of the Royal, Linnean, and Royal Meteorological
Societies; Honorary Member of the Royal Agricultural Society of England, of the
Chemico-Agricultural Society of Ulster, and of the Academy of Agriculture and
Forestry of Petrovskoie; Corresponding Member of the Institute of France
(Academy of Sciences), of the Society of Agriculturists of France,
and of the Agricultural Institute of Gorigoretsk; Chevalier
du Mérite Agricole (France); and Gold Medallist of
Merit for Agriculture (Emperor of Germany).*

LONDON: HENRY FROWDE,

OXFORD UNIVERSITY PRESS WAREHOUSE, AMEN CORNER, E.C.

1884.

INAUGURAL LECTURE.

OXFORD, MAY 6, 1884.

MR. VICE-CHANCELLOR AND GENTLEMEN,

THE term *Rural Economy*, which gives the title to
the Professorship to which I have been appointed,
might be held to include such a variety of topics, that
if I had no other guide in the selection of subjects to
bring before you, I should be placed in a position of
some difficulty. It would not be inappropriate to
treat of the general management of landed property
from the point of view of the Land-owner or the
Land-agent ; of the practical details of farming,
mechanical, economical, and commercial, such as are
essential to be observed by the tenant or occupier, if
his business is to be a profitable one ; or, lastly,
attention might be confined to the elucidation of the
scientific principles involved in successful practice, so
far as the existing knowledge of the day permitted.

Fortunately, however, I have not only the obvious
intentions of the Founder of this Chair, the late
Professor Sibthorp, and the way in which my Predecessor, the late Professor Daubeny, interpreted and
performed the duties of the office, to guide and limit
my selection of subjects ; but, on the 14th of July last,
these duties were, under the sanction of the Chancery
Division of the High Court of Justice, defined to be
to lecture on the *Scientific Principles of Agriculture.*

It may be not out of place to remind you that,
according to Dr. Sibthorp's will, the Sherardian Professor of Botany for the time being was to hold the
Professorship of Rural Economy ; and that hitherto,

B

that is until the settlement by the Court of Chancery last year, the appointment has been so held ; and the obligation was limited to giving one Public Lecture in each term. The new arrangement, however, dissociates the Chair of Rural Economy from that of Botany, and requires that twelve lectures shall be given annually.

Thus my responsibilities are somewhat defined, and they are, at the same time, somewhat limited :—for, how far from a complete system of instruction in all that pertains to Rural Economy can be embodied in a course of twelve lectures may be judged by reference to the course of study laid down in Institutions specially devoted to the training of Students in Agriculture. Thus, the Syllabus of one Agricultural College includes no less than thirty distinct subjects of lecture or other instruction, and the Prospectus of another enumerates sixteen main departments, covering much the same range of subjects as the other ; whilst each Institution has a numerous staff of Professors or Teachers. Again, in one London College a course of forty lectures is annually given on Soils, Manures, Crops, Live Stock, &c.; and in another lectures and laboratory instruction are given on Soils, Rotation, Manures, Feeding and Feeding-stuffs, Seeds, &c.

It is not then, I take it, expected of me that I should undertake to impart that systematic instruction in the various branches of the art and the science of Agriculture, which it is desirable the student who intends to devote himself to the details of practical farming, whether as agent or occupier, should receive. I assume that those studying at this University will for the most part be interested in land either as Owners, or as Clergymen in rural districts, or it may be as Statesmen, and that it should be my endeavour so to inculcate general principles as to train the observation, and give a direction to the reflections, of those who hear me, and to enable, and it may be to stimulate them, to study further for themselves, as problems requiring it may come before them.

What then is Agriculture—the scientific principles of which we have to investigate? Discarding the mere etymological meaning of the term, and giving it the'wider signification accorded to it by long usage, it is the art by which animal as well as vegetable products are obtained from the land.

Natural vegetation, as in the case of prairie-lands, for example, does indeed contribute food to the lower animals, and through them food and other products for the use of man ; but the result is attained with little of his aid which can come under the definition of Agriculture.

But the practices by which vegetable and animal products are obtained from the land in settled countries differ essentially from those above alluded to, and it is these to which our attention must mainly be directed. Such Agriculture implies the growth of other plants than would be obtained under conditions of natural vegetation, and the growth of more produce over a given area than would be so obtained. It implies, in fact, what may be called concentrated production.

Agriculture, the oldest of the arts, was in methods at first purely mechanical ; though, as we now know, the results to be attained were largely chemical. When manure was first applied to the soil, that is when the refuse of previous crops, or the excrements of animals, were first returned to the land as a means of increasing the growth of succeeding crops, directly chemical means—the supply of material—were first adopted.

Professor Daubeny, in his lectures on Roman Agriculture, quotes Cato as having said—

'If I am asked, what is the first point in good husbandry, I answer, good ploughing ; what the second, ploughing of any kind ; what the third, manuring.'

We have thus evidence of the relatively little esteem in which manuring was held in Italy two thousand years ago ; and in the fact that so much more value was set upon the mechanical operations we have evidence that the resources of the soil itself were far from being exhausted, and only required such means to be taken to render them available.

There is nevertheless evidence that, long before the time of Cato, it was sought to restore fertility to the soil by practices the efficacy of which is still fully recognised ; though the explanation of some of the phenomena involved is still a matter of controversy. Such, for example, was the growth of various crops of the Leguminous family, sometimes to be in great part removed, but in others to be at once ploughed into the land, by either of which methods the growth of succeeding crops was enhanced. This subject will receive detailed consideration further on.

But although manuring, in various ways, has thus been so long recognised as desirable, it is in fact only in quite recent times that the rationale of such practices could be at all satisfactorily explained. To this end it was obviously essential not only to know the composition of the vegetable products grown, but something of the sources of their various constituents —whether these must be derived from the soil, or whether from the atmosphere, or from water ?

What then is the composition, and what are the sources of the constituents of vegetable products ?

When a vegetable (or animal) substance is burnt,

the greater part of it is dissipated, but there remains a white ash. This ash is found on analysis to contain most or all of the constituents enumerated in the first column of the following Table.

Constituents of Plants and Animals.

INCOMBUSTIBLE OR FIXED.	COMBUSTIBLE OR VOLATILE.
Iron.	Carbon.
(Manganese).	Hydrogen.
Lime.	Oxygen.
Magnesia.	Nitrogen.
Potash.	
Soda.	
Phosphoric acid.	
Sulphuric acid.	.
Chlorine.	
Silica.	

Sulphur.
Phosphorus.

Rarer substances than these are also sometimes found. It will be my duty, in subsequent lectures, to call your attention in some detail to what is known as to the occurrence, and the offices, of the various mineral or ash-constituents in plants. It will suffice here to say, that the ash of one and the same description of plant, growing on different soils, may, so long as it is in the vegetative or immature state, differ very much in composition. Again, the ashes of different species, but growing on the same soil, will differ very widely in the proportion of their several constituents. But it is found that the nearer we approach to the elaboration of the final products of the plant—the seed for example—the more fixed is the composition of the ash of such products of one and the same species. In other words, there is very

little variation in the composition of the ash of one
and the same description of seed, or other final product, provided it be evenly and perfectly matured.
This fact alone, independently of what has been
established of late years in regard to the office or func-
tion so to speak of individual mineral constituents of
plants, would be sufficient to indicate the essentialness
of such constituents for healthy growth.

Th. De Saussure, in his work entitled, ' Recherches
sur la Végétation;' published in 1804, gave the results of the analyses of many plant-ashes, maintained
the essentialness of the ash-constituents, and pointed
out that they must be derived from the soil. He
also called attention to the probability that the
incombustible constituents so derived by plants from
the soil were the source of those found in the
animals which fed upon them.

Yet such was the prevailing uncertainty on the
point, that Sir Humphrey Davy, in his lectures
delivered not long afterwards, deemed it not inappropriate to combat the idea that the earths found
in plants had been formed from any of the elements
existing in the air, or in water. After quoting the
results of an experiment of his own, in which he
attempted to grow oats without any supply of silica
beyond that contained in the seed sown, and referring
to the experiments of De Saussure, he says :—

' As the evidence on the subject now stands, it
seems fair to conclude, that the different earths and
saline substances found in the organs of plants are
supplied by the soils in which they grow ; and in no
cases composed by new arrangements of the elements
in air or water.'

It is no longer doubted that the mineral or ash-
constituents of plants must be provided within the

soil, either from its own natural resources, or by means of manure. It will be seen, however, from the facts just stated, how very recently fixed ideas on the subject have been arrived at.

Then, as to the combustible or volatile constituents which are expelled in the incineration—the carbon, the hydrogen, the oxygen, and the nitrogen. From what we now know of the sources of these constituents of plants, it is obvious that a knowledge of the composition of the atmosphere and of water was essential to any true conception of the main features of the vegetative process, and it was only towards the end of the last century that the composition of the air, and of water, and their mutual relations with vegetation, were first pointed out.

I shall have to go into these matters in some detail in subsequent lectures, but I would here observe that it is to the collective labours of Black, Scheele, Priestley, Lavoisier, Cavendish, and Watt, that we owe the knowledge that common air consists of nitrogen and of oxygen, with a little carbonic acid ; that carbonic acid itself is composed of carbon and of oxygen; and that water is composed of hydrogen and oxygen ; whilst Priestley and Ingenhousz, Sennebier and Woodhouse, investigated the mutual relations of these bodies and vegetable growth.

Thus, Priestley observed that plants possessed the faculty of purifying air vitiated by combustion, or by the respiration of animals; and he having discovered oxygen, it was found that the bubbles which Bonnet had shown to be emitted from the surface of leaves immersed in water consisted chiefly of that gas. Ingenhousz demonstrated that the action of light was essential to the development of these phenomena, and Sennebier proved that the oxygen evolved resulted from the decomposition of the carbonic acid taken up.

So far, attention had been more prominently directed to the influence of plants upon the media with which they were surrounded, than to that of those media in contributing to the increased substance of the plants themselves.

It was, too, towards the end of the last century, and in the beginning of the present one, that De Saussure followed up these enquiries; and, in his work already alluded to, he may be said to have indicated, if not indeed established, some of the most important facts with which we are yet acquainted regarding the sources of the constituents of the growing plant. He illustrated experimentally, and even to a certain extent quantitatively, the fact that in sun-light plants increase in carbon, hydrogen, and oxygen, at the expense of carbonic acid and of water; and in his main experiment on the point he found the increase in carbon, and in the elements of water, was very closely in the proportion in which they are known to exist in the chief non-nitrogenous constituents of plants, the carbohydrates—starch, gum, sugar, cellulose, etc.

With regard to the nitrogen which plants had already been shown to contain, Priestley and Ingenhousz thought their experiments indicated that they absorbed free nitrogen from the atmosphere; but Sennebier and Woodhouse arrived at an opposite conclusion. De Saussure, again, thought that his experiments showed rather an evolution of nitrogen at the expense of the substance of the plant than any assimilation of it from gaseous media. He further concluded that the source of the nitrogen of plants was more probably the nitrogenous compounds within the soil, and the small amount of ammonia which he demonstrated to exist in the atmosphere.

Upon the whole, De Saussure concluded that air and water contributed a much larger proportion of the dry substance of plants than did the soils in which they grew. In his view a fertile soil was one which yielded liberally to the plant nitrogenous compounds, and the incombustible or mineral constituents; whilst the carbon, hydrogen, and oxygen, of which the greater proportion of the dry substance of the plant was made up, were at least mainly derived from the air and water.

From what has been stated, it will be seen that anything like a consistent scientific explanation of vegetable and animal production was impossible until comparatively very recent times. In fact it is clear that agricultural chemistry properly so called is not a century old; whilst, without a knowledge of it, the scientific principles of agriculture could not be taught.

What then is agricultural chemistry? It is the chemistry of the soil, the chemistry of the atmosphere, the chemistry of vegetation, and the chemistry of animal life and growth, in their bearings on agricultural production.

Obviously vegetable and animal physiology are essential elements in the explanation of many of the phenomena of agricultural production; but for anything more than necessary passing reference to these branches of science you must look to the distinguished Professors of Physiology and Botany now attached to this University.

To Sir Humphrey Davy we owe the first systematic attempt to apply scientific principles to the elucidation and improvement of agricultural practice.

Each year, for ten successive years, from 1802 to
1812, he delivered a course of lectures which, in
1813, were published in a work entitled, 'Elements
of Agricultural Chemistry,' which he finally revised
for the fourth edition in 1827, but which has gone
through several editions since.

In those lectures Sir Humphrey Davy passed in
review and correlated the then existing knowledge,
both practical and scientific, bearing upon agriculture.
He treated of the influences of heat and light; of the
organisation of plants; of the difference, and the
change, in the chemical composition of their different
parts; of the sources, composition, and treatment of
soils; of the composition of the atmosphere, and its
influence on vegetation; of the composition and the
action of manures; of fermentation and putrefaction;
and, finally, of the principles involved in various
recognised agricultural practices.

Concurrently with the delivery and the publication
of Davy's Lectures in England, the most prominent
writer on the scientific principles of agriculture on
the continent was Thaer. In 1810 he published a
work entitled 'Principes Raisonnés d'Agriculture,'
and for some years afterwards contributed papers on
some special points. He considered that the fertility
of the soil depended on the amount and character
of the humus it contained; it being, in his view, the
only substance, excepting water, that yielded aliment
to plants. He pointed out that it was the residue
of previous vegetation, containing carbon, hydrogen,
oxygen, and nitrogen, associated with phosphorus,
sulphur, and some salts. He quotes De Saussure
as having shown that humus contains a lower percentage of oxygen and higher percentages of carbon
and nitrogen than the plants from which it has been

produced. He states that its composition varies according to the access of oxygen and water; that it absorbs oxygen, and gives out carbonic acid, which supplies nourishment to plants; also, that it yields soluble extractive matter again and again by time and exposure. He further states that the more the decomposition has proceeded, the more refractory, and the less active and useful for vegetation, is the residue.

With the exception of the discourses of Sir Humphrey Davy, and the writings of Thaer, the subject seems to have received comparatively little attention, nor was any other addition of importance made to our knowledge in regard to it, during a period of more than twenty years, from the date of the appearance of De Saussure's work in 1804.

From about 1825 to 1840, Dr. Carl Sprengel, formerly Professor of Agriculture at Brunswick, published a series of about thirty papers in connection with agriculture, and agricultural chemistry. These memoirs covered a wide range of subjects, and recorded the results of numerous investigations by the author himself. Among them may be mentioned investigations on humus, humic acid, and humates; on the constituents of some surface soils and subsoils; on the composition of various kinds of straw, and their value as food and litter; on the amount of potash in granites, and other rocks; on the ash-constituents of cereal grains, &c. Lastly, he published a general treatise on manures, including a chapter on animal manures, in which he gives the results of the analysis of the solid and liquid excretions of the various animals of the farm, and, among other points, insists upon the value as manure of the ammonia yielded by such materials.

Professor Schübler, of Tübingen, also published

a Work entitled, 'Grundsätze der Agricultur-Chemie,' a second and revised edition of which was brought out in 1838, by Professor Krutzsch, of Tharand. In this Work the physical and chemical properties of soils are discussed in much detail; and the results of numerous investigations on the subject by the author himself are given.

Boussingault, who had previously published numerous papers, chiefly on chemical subjects, about 1834 became, by marriage, joint proprietor with his brother-in-law of the estate of Bechelbronn, in Alsace. His brother-in-law, M. Lebel, was both a chemical manufacturer and an intelligent practical farmer, accustomed to use the balance for the weighing of manure, crops, and cattle. Boussingault at once applied himself to Chemico-Agricultural research; and it was under these conditions of the association of 'Practice with Science' that the first laboratory on a farm was established.

From this time forward, Boussingault generally spent about half the year in Paris, and the other half in Alsace; and he has continued his scientific labours, sometimes in the city and sometimes in the country, up to the present time. I may here mention that I had the pleasure of seeing him well, and still actively interested in problems of agricultural science, at his Chateau in Alsace in the autumn of last year.

Boussingault's first important contribution to agricultural chemistry was made in 1836, when he published a paper on the amount of nitrogen in different foods, and on the equivalence of the foods, founded on the amounts of nitrogen they contained; and he compared the results so arrived at with the estimates of others founded on actual experience.

Although his conclusions on the subject have doubtless undergone modification since that time, the work itself marked a great advance on previously existing knowledge and modes of viewing the question.

In 1837 Boussingault published papers on the amount of gluten in different kinds of wheat; on the influence of the clearing of forests on the diminution of the flow of rivers; and on the meteorological influences affecting the culture of the vine. In 1838 he published the results of an elaborate research on the principles underlying the value of a rotation of crops. He determined by analysis the composition, organic and inorganic, of both the manures applied to the land, and the crops harvested. In his treatment of the subject he evinced a clear perception of the most important problems involved in such an enquiry; some of which, with the united labours of himself and many other workers, have scarcely yet received an undisputed solution.

Thus, in the same year (1838), he published the results of an investigation on the question whether plants assimilate the free or uncombined nitrogen of the atmosphere; and although the analytical methods of the day were inadequate to the decisive settlement of the point, his conclusions were in the main those which much subsequent work of his own, and much of others also, have served to confirm.

In subsequent lectures I shall have to refer in some detail to the various investigations here alluded to.

As a further element of the question of the chemical statistics of rotation, Boussingault determined the amount and composition of the residues of various crops; also the amount of constituents consumed in the food of a cow, and of a horse, respectively, and yielded in the milk and excretions of the cow, and in the excretions of the horse. Here, again, the exigencies of the investigation he undertook were beyond the reach of the known chemical methods of the time. Indeed, rude as the art of agriculture is generally considered to be, the scientific elucidation of its practices requires the most refined and very varied methods of research ; and a characteristic of the work, not only of Boussingault, but of most agricultural investigators, may be said to be, that they have frequently had to devise methods suitable to their purpose, before they could grapple with the problems before them.

In 1839, chiefly in recognition of his important contributions to agricultural chemistry, Boussingault was elected a member of the Institute of France ; and in 1878, thirty-nine years later, the Council of the Royal Society awarded to him the Copley Medal, the highest honour at their disposal, for his numerous and various contributions to science, but especially for those relating to agriculture.

The foregoing brief historical sketch is sufficient to indicate, though but in broad outline, the range of existing knowledge on the subject of agricultural chemistry prior to the appearance of Liebig's memorable work, 'Organic Chemistry in its applications to Agriculture and Physiology,' the first edition of which was published in 1840.

It will be seen that some very important, and indeed fundamental, facts had already been established in regard to vegetation; first by the numerous investigations made about half a century previously, by which the composition of the atmosphere, and of water, and the mutual relations of these and vegetable growth,

had been determined; next by the labours of De
Saussure, to which we owe a near approach to a quantitative representation of the phenomena of vegetation, which had thus far only been qualitatively
observed. To De Saussure we also owe a clear perception of the importance, and of the sources, of both
the nitrogen, and the ash-constituents, of plants. Then
followed Davy, with the first attempt to give a systematic view of the relations of practice with science
in agriculture; Thaer, who traced the fertility of
soils to the residue of previous vegetation which they
contained; Sprengel, who contributed much experimental result on various branches of the subject; and
lastly Boussingault, who had not only still further
extended experimental enquiry, but brought both
his own and previous results to bear upon the elucidation of long-recognised agricultural practices.

There can be no doubt that the data supplied by
the researches which have been referred to, and especially those of De Saussure, Davy, Thaer, Sprengel,
and Boussingault, contributed important elements to
the basis of established facts upon which Liebig
founded his brilliant generalisations. Indeed, so obvious was this, that, in 1841, Dumas and Boussingault
published, jointly, an essay which afterwards appeared
in English under the title of 'The Chemical and
Physiological Balance of Organic Nature,' which was,
in fact, a sort of 'réclamation.'

Nor can there be any doubt that the appearance
of Liebig's two works, 'Organic Chemistry in its
applications to Agriculture and Physiology' in 1840,
and 'Animal Chemistry, or Organic Chemistry in its
applications to Physiology and Pathology' in 1842,
constituted a very marked epoch in the history of
the progress of Agricultural Chemistry. In the

treatment of his subject he not only called to his aid
the previously existing knowledge directly bearing
upon it, but he also turned to good account the more
recent triumphs of organic chemistry, many of which
had been won in his own laboratory. Further, a
marked feature of his expositions was the adoption of
the quantitative method of illustration, and argument.

But, notwithstanding the evidence afforded by the
direct experiments of De Saussure and his prede-
cessors, Vegetable Physiologists, and some others, continued to hold the view that the humus of the soil
was the source of the carbon of vegetation. Liebig
gave full weight to the evidence of the experiments
of De Saussure and others; he illustrated the possible or probable transformations within the plant
by facts already established in organic chemistry;
and he demonstrated the impossibility of the humus
of the soil supplying the amount of carbon assimilated over a given area. He pointed out, that
humus itself was the product of previous vegetable
growth; that it could not therefore be an original
source of the carbon; and that, from the degree of
its insolubility, either in pure water or in water containing alkaline or earthy bases, only a small portion
of the carbon assimilated by plants could be derived
from the amount of humus that could possibly enter
the plant in solution. He maintained that, so far as
humus was beneficial to vegetation at all, it was
only by its oxidation, and a consequent supply of
carbonic acid within the soil; a source which he
considered only of importance in the early stages
of the life of a plant, and before it had developed,
and exposed to the atmosphere, a sufficient amount
of green surface to render it independent of soil-
supplies of carbonic acid.

In subsequent lectures, evidence will be adduced showing that Liebig was certainly right, in concluding that humus is not an adequate source of the carbon of our crops. It will be seen that, to some of them at any rate, it is at most only a very limited source, if indeed it is to them a source at all. It will, on the other hand, be shown that the organic residue of previous vegetation accumulated in the soil is, to say the least, a very material source of the nitrogen of our crops.

Thus, though mistaken as to the explanation of the fertility of soils rich in humus, Thaer and others were, after all, not far from the truth when they maintained that the richness of a soil in such matter in a condition readily susceptible to oxidation, was, in a great degree, the measure of its fertility.

With regard to the hydrogen of plants, at any rate that portion of it contained in their non-nitrogenous products, Liebig maintained that its source must be water; and that the source of the oxygen was either that contained in carbonic acid, or that in water.

With regard to the nitrogen of vegetation, both from the known characters of free nitrogen, and as he considered a legitimate deduction from direct experiments, Liebig argued that plants did not assimilate uncombined nitrogen, either from the atmosphere, or dissolved in water and so absorbed by the roots. The source of the nitrogen of vegetation was, he maintained, *ammonia*; the product of the putrefaction of one generation of plants and animals affording a supply for its successors. He pointed out that, in the case of a farm receiving nothing from external sources, and selling off certain products, the amount of nitrogen in the manure produced by the consumption of some of the vegetable produce on

the farm itself, together with that due to the refuse
of the crops, must always be less than was contained
in the crops grown. He concluded that, although
the quantity returned to the land as manure was
important, a main source of the nitrogen assimilated
over a given area was the ammonia brought down
from the atmosphere in rain.

There can be no doubt that, owing to the limited
and defective experimental evidence then at command
on the point, Liebig at that time (as he did afterwards)
greatly over-estimated the amount of combined nitrogen available to vegetation from that source.

In Boussingault's 'réclamation' already referred
to, he gave much more prominence to the importance
of the nitrogen of manures.   In Liebig's next edition
(in 1843) he combated the notion of the relative
importance of the nitrogen of manures.   He maintained, in opposition to the view put forward in his
former edition, that the atmosphere afforded a
sufficient supply of combined nitrogen not only for
natural vegetation, but for cultivated plants ; that
this supply was sufficient for the cereals as well as
for leguminous plants ; that it was not necessary to
supply nitrogen for the former ; and he insisted very
much more strongly than formerly, on the relative
importance of the supply of the incombustible, or,
as he designated them, the 'inorganic' or 'mineral,'
constituents.

Many determinations of the amount of combined
nitrogen brought down in rain have been made, in
different countries, since the date here referred to ;
and in a subsequent lecture it will be shown that
the amount so available to the vegetation of a given
area is very much less than was assumed by Liebig,
or has generally been supposed.

As to the incombustible, or, as he designated them, the '*mineral*' constituents, Liebig adduced many illustrations in proof of their essentialness. He called attention to the variation in the composition of the ash of plants grown on different soils ; and he assumed a greater degree of mutual replaceability of one base by another, or of one acid by another, than could now be admitted. He traced the difference in the mineral composition of different soils to that of the rocks which had been their source ; and he seems to have been led by the consideration of the gradual action of '*weathering,*' in rendering available the otherwise locked-up stores, to attribute the benefits of *fallow* exclusively to the increased supply of the incombustible or mineral constituents which would, by its agency, be brought into a condition in which they could be taken up by plants.

It will be seen further on, how very materially subsequently acquired experimental evidence has tended to modify our views as to the explanation of the benefits of fallow.

The benefits of an alternation of crops, Liebig considered to be in part explained by the influence of the excreted matters from one description of crop upon the growth of another. He did not attach weight to the assumption that such matters would be directly injurious to the same description of crop; but he supposed rather that the matters excreted by a plant were those which it did not need, and that they would therefore be of no avail to the same description of plant, but would be of use to others. He, however, attributed much of the benefits of a rotation, to different mineral constituents being required from the soil by the different crops.

Since the enunciation of these views, very much

direct experimental evidence bearing upon the rationale of rotation has been acquired, and it will be my duty to lay it before you in some detail.

Treating of *manure*, Liebig laid the greatest stress on the return by it of the potass and the phosphates removed by the crops. But, in his first edition, he also insisted on the importance of the nitrogen, especially that in the liquid excretions of animals, and condemned the methods of treatment of animal manures by which the ammonia was allowed to be lost by evaporation. It is curious, and significant, however, that some of the passages in that edition, in which he the most forcibly urges the value of the nitrogen of animal manures, are omitted in the third and fourth editions.

In his second work, that on *Animal Chemistry*, published in 1842, Liebig discussed two subjects of much interest and practical importance to the agriculturist :—namely, the sources in the food of the fat stored up in the animal body, and the characteristic food requirements of the animal organism induced by the exercise of force.

To render the points here at issue intelligible, it is necessary to remind you that the constituents of food, both vegetable and animal, may broadly be classed into those containing nitrogen, and those not containing it—in other words, into the nitrogenous, and the non-nitrogenous constituents.

From the nitrogenous constituents of the food, the nitrogenous constituents of the animal body—the membranes and cellular tissue, the nerves and brain, cartilage, and the organic part of bones—must be derived. It is admitted that by oxidation and transformation within the body, some of these nitrogenous matters may yield fat, with other products, such as urea which passes off by the urine, and carbonic acid which passes off by respiration.

The non-nitrogenous constituents of the food on the other hand—the fatty matters, and the so-called carbohydrates—starch, sugar, cellulose, &c., are supposed to be the chief sources of the carbonic acid respired, and hence have sometimes been classed as the respiratory constituents of food.

The questions arise—Is the non-nitrogenous substance *fat*, which is stored up in the feeding of the herbivora, derived chiefly from the nitrogenous matters of the organs, or in the circulation, or from the non-nitrogenous matters of the food, the carbohydrates? Then, again, in the exercise of force, is the increased amount of the products of oxidation eliminated from the body chiefly due to an increased transformation of the nitrogenous substance of the tissues and fluids, or of the non-nitrogenous constituents—the fat of the body, or of the food, and the carbohydrates?

In reference to the sources of the fat of the animal body, Liebig maintained that the vegetable food consumed by herbivora did not contain anything like the amount of fat which they stored up in their bodies; and he showed how nearly the composition of fat was obtained by the simple elimination of so much oxygen, or of oxygen and a little carbonic acid, from the various carbohydrates of the vegetable food—starch, sugar, &c. Much less oxygen would be required to be eliminated from the nitrogenous constituents, such as fibrine, &c., than from a quantity of carbohydrate containing an equal amount of carbon. The formation of fatty matter in plants was of the same kind; it was the result of a secondary action, starch being first formed from carbonic acid and water.

Referring to the exercise of force, be argued that the animal secretions must contain the products of the metamorphosis of the tissues; he concluded that a starving man, with severe exertion, would secrete more urea than the most highly fed individual in a state of rest; and he combated the idea that the nitrogen of the food can pass into urea without having previously become part of an organised tissue. He said :—

'As an immediate effect of the manifestation of mechanical force, we see that a part of the muscular substance loses its vital properties, its character of life; that this portion separates from the living part, and loses its capacity of growth and its power of resistance. We find that this change of properties is accompanied by the entrance of a foreign body (oxygen) into the composition of the muscular fibre * * *; and all experience proves, that this conversion of living muscular fibre into compounds destitute of vitality is accelerated or retarded according to the amount of force employed to produce motion. Nay, it may safely be affirmed, that they are mutually proportional; that a rapid transformation of muscular fibre, or, as it may be called, a rapid change of matter, determines a greater amount of mechanical force; and conversely, that a greater amount of mechanical motion (of mechanical force expended in motion) determines a more rapid change of matter.'

Again :—

'The change of matter, the manifestation of mechanical force, and the absorption of oxygen, are, in the animal body, so closely connected with each other, that we may consider the amount of motion, and the quantity of living tissue transformed, as proportional to the quantity of oxygen inspired and consumed in a given time by the animal.'

And again:—

'The production of heat and the change of matter are closely related to each other; but although heat can be produced in the body without any change of matter in living tissues, yet the change of matter cannot be supposed to take place without the co-operation of oxygen.'

Further, on the same point:—

'The sum of force available for mechanical purposes must be equal to the sum of the vital forces of all tissues adapted to the change of matter.'

'If, in equal times, unequal quantities of oxygen are consumed, the result is obvious, in an unequal amount of heat liberated, and of mechanical force.'

'When unequal amounts of mechanical force are expended this determines the absorption of corresponding and unequal quantities of oxygen.'

Then, more definitely still, referring to the changes which take place coincidently with the exercise of force, and to the demands of the system for repair accordingly, he says:—

'The amount of azotised food necessary to restore the equilibrium between waste and supply is directly proportional to the amount of tissues metamorphosed.'

'The amount of living matter, which in the body loses the condition of life, is, in equal temperatures, directly proportional to the mechanical effects produced in a given time.'

'The amount of tissue metamorphosed in a given time may be measured by the quantity of nitrogen in the urine.'

'The sum of the mechanical effects produced in

two individuals, in the same temperature, is proportional to the amount of nitrogen in their urine; whether the mechanical force has been employed in voluntary or involuntary motions, whether it has been consumed by the limbs or by the heart and other viscera.'

Thus, apparently influenced by the physiological considerations which he had adduced, and notwithstanding that in some passages he seemed to recognise a connection between the total quantity of oxygen inspired and consumed, and the quantity of mechanical force developed, Liebig nevertheless very prominently insisted that the amount of muscular tissue transformed—the amount of nitrogenous substance oxidated—was the measure of the force generated. He accordingly draws the conclusion, that the requirement for the nitrogenous constituents of food will be increased in proportion to the increase in the amount of force expended.

It will be obvious that the question whether, in the feeding of animals for the exercise of mechanical force, that is, for their labour, the demands of the system will be proportionally the greater for the nitrogenous, or for the non-nitrogenous constituents of food, is one of considerable interest and practical importance.

In reference to this point, to that of the sources in the food of the fat of the animal body, as well as to the requirements for the different constituents of food for the maintenance, and for the general increase, of the body, in the feeding of the animals of the farm, a great deal of experimental evidence has been acquired during the last forty years, both in this country and on the Continent, and to this I shall have to refer in some detail on a future occasion.

So far, I have endeavoured to convey some idea of the state of knowledge bearing upon the scientific principles of agriculture, prior to the appearance of Liebig's first two works on the subject; and also briefly to summarise the views he then enunciated in regard to some points of chief importance. I will now attempt briefly to indicate what progress has been made since that period, largely at any rate due either to the direct influence of his teaching, or to the stimulus given to enquiry by the discussions which his writings called forth.

It is a coincidence of some interest, that the first lectures given by the Sibthorpian Professor of Rural Economy in this University, were delivered by my predecessor, the late Dr. Daubeny, almost contemporaneously with the appearance of Liebig's first Work, in 1840; one of a course of three having been given before, and two within a few months after, the publication of that work. These lectures were afterwards published, and, in his preface, Professor Daubeny expresses his indebtedness to Liebig's Work for some of the fundamental doctrines, and for some of the details embodied in his expositions.

In the lectures in question, Professor Daubeny discussed the importance of the study of botany, physiology, and chemistry, in the elucidation of agricultural practices. He contrasted the conditions of cultivation in the case of virgin soils, with those in long-settled countries. He treated of the sources of the constituents of our crops, of fallowing, manuring, and rotation, so far as the knowledge of the day permitted.

Finally, Dr. Daubeny put forward some speculations as to the origin of the constituents of the first vegetation on the surface of the globe, and especially as to the source of the combined carbon, and the combined nitrogen, accumulated in the total existing vegetable and animal life and remains. He traced these to carbonic acid, and ammonia, evolved by volcanic action. It may be observed that one source of combined nitrogen is undoubtedly electrical action, especially in equatorial regions, but it will be seen further on, that so far as quantitative evidence is at command on the point, the amounts of combined nitrogen available from atmospheric sources over a given area, within a given time, are, at any rate in temperate latitudes, quite inadequate to account for the amounts recovered in crops over the same area, and in the same time. It will be seen too, that the question of the sources of the nitrogen of our crops, is one upon which very conflicting views are still entertained.

For a period of more than twenty-five years, Dr. Daubeny continued to give lectures from time to time on various branches of Rural Economy, generally with some reference to the discussions of the day.

But independently of discourses of this expository or critical kind, he, in 1845, described the results of some experiments he had made to elucidate the principles involved in the rotation of crops. To this end he had set apart a number of plots in the Botanic Garden, each 10 feet square. On some of these he grew the same description of plant year after year for several years in succession, whilst on others he alternated the different crops. The soil was admittedly not very favourable, being made ground, and the plots were very small. Still the results clearly showed that more produce was obtained from a given area when the different plants were grown in alternation, than when . the same description was grown successively. Subsequent examination of the soils further showed considerable differences, dependent on the exhaustion by the different crops.

Dr. Daubeny confined his illustrations and his discussions almost entirely to the mineral or ash-constituents of the crops; thus following the lines of argument current at the time. Although the facts brought out are of interest, all subsequently acquired evidence tends to show, that the benefits of rotation are not explicable by exclusive reference to the difference in the description, and amounts, of the mineral constituents which are taken up by the different crops.

Among the most notable of Professor Daubeny's lectures, was a course of eight on Roman Husbandry, which were afterwards published.

These lectures evince considerable historical research, and are of much interest independently of the facts relating to the Agriculture of the Ancients which they bring to light. But their chief interest to the agricultural student of the present day is to be found in the evidence, discernible between the lines, that certain practices then adopted for the increase of the products obtained from the land, are still recognised as effective, though, in some cases, the precise explanation of the benefits derived yet remains a matter of controversy. Thus, although no fixed rotations of crops seem to have been adopted, yet the occasional growth of plants of the Leguminous family was had recourse to, and recognised as a means of increasing the growth of the gramineous crops with which they were alternated.

Again, the evidence of both Columella and Virgil goes to show, that the fertility of soils was then attributed to the accumulations from previous natural vegetation, and that, as this store was gradually used up, the soils became poor.

Although, as has been said, the lectures which have been given by the Sibthorpian Professor of Rural Economy commenced almost simultaneously with the appearance of Liebig's first work, and many of them had direct reference either to Liebig's own views, or to current discussions relating to them, their inauguration was prior to, and quite independent of, the interest excited by that work. Not so was it, in the case of many other agencies for promoting knowledge in regard to the scientific principles of agriculture, both in this country, and on the Continents of Europe and America. I think too, I may safely say, that the stimulus was earlier felt, and was earlier productive of results, in this country, than in Liebig's own, or elsewhere.

In 1843, the Royal Agricultural Society of England first appointed a consulting Chemist; Dr., now Sir Lyon Playfair, being the first holder of the office. In 1848, the late Professor Way was elected, and the Society's journals of that time bear testimony to his clear perception of the agricultural problems requiring solution, and of his capacity as an investigator. In 1858, Dr. Voelcker succeeded to the office, and continues to hold it with much advantage to that union of 'Practice with Science,' which the Society by its motto recognises as so essential to progress. Thus, after having been Professor of Chemistry at the Royal Agricultural College, Cirencester, for many years, Dr. Voelcker has now been consulting Chemist to the Royal Agricultural Society of England for more than a quarter of a century; and to some of the results of his investigations I shall have to refer on future occasions.

It was also in 1843, that there was established the Chemico-Agricultural Society of Scotland, which was, I believe, broken up after it had existed between four and five years, because its able Chemist, the late Professor Johnston, failed to find a remedy for the potato disease. Somewhat similar duties, including a good deal of agricultural research, have however since been performed under the auspices of the Highland and Agricultural Society of Scotland, for many years by the late Professor Anderson, and more recently by Dr. Aitken.

In 1845, the Chemico-Agricultural Society of Ulster was established ; Professor Hodges was appointed as Chemist, and he continues ably to perform the duties of the office.

As already intimated, agencies of this kind were not so soon brought into operation on the Continent. Nevertheless, the numerous Agricultural Experimental Stations which have been established, not only in Germany, but in most continental States, owe their origin very directly to the writings, the teachings, and the influence of Liebig.

The movement seems to have originated in Saxony, where Stöckhardt had already stimulated interest in the subject by his lectures and his writings. After some correspondence in 1850–51, between the late Dr. Crucius and others on the one side, and the Government on the other, the first so-called 'Agricultural Experimental Station' was established at Möchern, near Leipzig, in 1851–2. In 1877 the twenty-fifth anniversary of the foundation of that Institution was celebrated at Leipzig ; when an account, which has since been published, was given of the number of Stations then existing, of the number of Chemists engaged, and of the subjects which had

been investigated. From that statement we learn
that at that period the number of Stations was :—

In the various German States . . .	74
In Austria	16
In Italy	10
In Sweden	7
In Denmark	1
In Russia	3
In Belgium	3
In Holland	2
In France	2
In Switzerland	3
In Spain	1
Total	122

Thus, seven years ago there were 122 'Agricultural
Stations' on the Continent of Europe, and the number has doubtless by this time considerably increased.

Each of these Stations is under the direction of a
Chemist, frequently with one or more assistants.
One special duty of most of them is to examine or
analyse, and report upon, manures, seeds, or feeding-
stuffs, offered for sale to the farmer ; and it seems to
have been found to be the interest of the dealers in such
commodities, to submit their proceedings to a certain
degree of supervision by the Chemist of the Station
of their District.

But agricultural research has also been a characteristic feature of these institutions.   It was stated in
the report referred to, that the investigation of soils
had been the prominent object at sixteen of them ;
experiments with manures at twenty-four ; vegetable
physiology at twenty-eight ; animal physiology and
feeding experiments at twenty ; vine culture and
wine making at thirteen ; forest culture at nine ; and
milk production at eleven.   Others, according to their
locality, have devoted special attention to fruit
culture, olive culture, the treatment of moor, bog and

peat land, the production of silk, the manufacture of spirit, and other products.

But few experimental stations having in the main similar objects to those of the Stations on the Continent of Europe, have been established either in our own Country or in America; but in the German Report above cited, Scotland is credited with one, and the United States with one.

There are, however, several Stations in the United States, where agricultural investigation is carried on; and a writer in that Country has recently made an appeal to the Government, to establish Stations with a view to the investigation of each general variety of soil and climate in the United States; to assign a Chemist to each, to institute experiments with the crops most suitable to the locality, to analyse the soil, &c., and to report the results.

The records of the results of the investigations conducted at the large number of experimental stations on the Continent of Europe, are extremely voluminous; and the number of systematic works which have appeared on various branches of the subject in the French and German languages during the last forty years is very great. I will here refer to the several editions of Liebig's first two works already alluded to, his 'Familiar Letters on Chemistry,' his 'Modern Agriculture,' his 'Principles,' and finally his 'Natural Laws of Husbandry,' all of which have appeared in the English language. Reference should also be made to Boussingault's 'Economie Rurale' the first edition of which was published in English in 1845. It has, however, gone through other editions in France; and subsequently, at intervals from 1860 to 1878, Boussingault published a series of six volumes, entitled—'Agronomie, Chimie Agricole et

Physiologie'—and covering a still wider range of subjects, bringing the information on the various points up to date, and going into much detail as to the methods of research, as well as to the results obtained. In these volumes, those who desire it will be able to examine for themselves the evidence upon which many important conclusions have been based.

A few of the works by English authors which have appeared within the same period, it may be well to enumerate. Among them are Johnston's 'Lectures on Agricultural Chemistry and Geology,' first published in 1844 ; and his 'Experimental Agriculture,' published in 1849. His successor, Dr. Anderson, also published, in 1860, a work entitled—'Elements of Agricultural Chemistry.'

New editions of Johnston's work have since been published by Dr. C. Cameron ; the first of these, the tenth of the original work, appearing in 1877.

Perhaps the most compendious record of the results of the Continental investigations up to the time of its publication, which has appeared in our own language, is that by Professor S. W. Johnson, of Yale College, Newhaven, Connecticut, in two volumes, entitled ' How Crops Grow,' and ' How Crops Feed.' The results of the German experiments on the Feeding of Animals have also been summarised by Dr. Armsby, in a volume published in America, and entitled—' Manual of Cattle-Feeding.'

Neither in the works above referred to, nor in the reports of the Continental Experimental Stations, do we find the record of many systematic or long continued field-experiments. We have, however, the example of Boussingault, and the opinions of Sir Humphrey Davy, Liebig, and Daubeny, as to the great importance of such methods of agricultural research.

Davy says:—

'Nothing is more wanting in agriculture than experiments in which all the circumstances are minutely and scientifically detailed. This art will advance with rapidity in proportion as it becomes exact in its methods. As in physical researches, all the causes should be considered; a difference in the results may be produced, even by the fall of a half-inch of rain more or less in the course of a season, or a few degrees of temperature, or even by a slight difference in the subsoil, or in the inclination of the land.'

'Information collected after views of distinct inquiry, would necessarily be fitted for inductive reasoning, and capable of being connected with the general principles of science; and a few histories of the results of truly philosophical experiments in agricultural chemistry would be of more value in enlightening and benefiting the farmer, than the greatest possible accumulation of imperfect trials, conducted merely in the empirical spirit.'

'It is from the higher classes of the community, from the proprietors of land,—those who are fitted by their education, to form enlightened plans, and by their fortunes, to carry such plans into execution; it is from these that the principles of improvement must flow to the labouring classes of the community; and in all classes the benefit is mutual; for the interest of the tenantry must be always likewise the interest of the proprietors of the soil.'

'Discoveries made in the cultivation of the earth are not merely for the time and country in which they are developed, but they may be considered as extending to future ages, and as ultimately tending to benefit the whole human race; as affording subsistence for generations yet to come; as multiplying life; and not only multiplying life, but likewise providing for its enjoyment.'

Liebig said :—' I shall be happy if I succeed in attracting the attention of men of science to subjects which so well merit to engage their talents and energies. Perfect agriculture is the true foundation of all trade and industry—it is the foundation of the riches of states. But a rational system of agriculture cannot be formed without the application of scientific principles ; for such a system must be based on an exact acquaintance with the means of nutrition of vegetables, and with the influence of soils and actions of manure upon them. This knowledge we must seek from chemistry, which teaches the mode of investigating the composition and of studying the characters of the different substances from which plants derive their nourishment.'

' Since the time of the immortal author of the " Agricultural Chemistry" (Davy) no chemist has occupied himself in studying the application of chemical principles to the growth of vegetables, and to organic processes. I have endeavoured to follow the path marked out by Sir Humphrey Davy, who based his conclusions only on that which was capable of inquiry and proof. This is the path of true philosophical inquiry, which promises to lead us to truth —the proper object of our research.'

The importance which Professor Daubeny attached to field-experiments, is evidenced in the facts that, at a very early period of his connection with the study and teaching of the principles of Rural Economy, he instituted the experiments on Rotation, in the Botanic Garden, already referred to ; and that subsequently, I believe in 1860, he bequeathed to his successors in the Chair a piece of ground of about one and a

half acre for the purposes of experiments; feeling, as
he said in a lecture on the subject, that the objects
of the foundation of the Professorship of Rural
Economy—' would not be fully attained until the
holder of it was enabled, not only to retail the information he might obtain from books, but also to
illustrate it by experiment, and to verify, as well
as extend the knowledge he might have derived from
others through original investigations of his own.'

Briefly described, the subjects which Dr. Daubeny
indicated as suitable for such enquiry were—1. To
determine what amount of mechanical treatment, and
what length of time, would be required to bring an
exhausted soil back to fertility without manure.
2. To try whether, if a soil be rich in mineral constituents, ammoniacal manures may not be dispensed
with by the use of mechanical operations; since, as
he says, Liebig's opinion is, that many plants have
the power 'of absorbing from the atmosphere so
large an amount of ammonia, as would seem to
render them independent of animal manures, and to
enable them to derive all their ingredients, except
their mineral ones, from the atmosphere.' 3. The
causes of the failure of clover. 4. The effects of
gypsum applied as manure. 5. To try whether there
is any disadvantage in the use of superphosphates
from which the fluorine has been dissipated, compared with bones from which it has not. 6. Whether
growing plants year after year on the same land
from the seed yielded, tends to variation, or to the
obliteration of specific characters. 7. Lastly, to determine how far the growth of the fungi which
attack different crops may be considered as a cause,
or only the effect, of disease.

I may here say that much experimental evidence

is now available with regard to some of these questions, and I shall have to call attention to it in
subsequent lectures.

I may add that, very shortly after Dr. Daubeny
had bequeathed the piece of ground to which I have
referred, I visited it with him, and at the time ex-
pressed a fear that it was not very suitable for the
objects in view, little thinking that it would fall to
my lot to consider the subject more seriously, as
I shall soon have to do. In the first place, I pointed
out the unlevel character of the plot, and gathered
from the history of it given to me, that it was very
uneven as to condition. Accordingly, I suggested
that, before any experiments on it were undertaken,
it should be brought into an even state as far as
mechanical operations could accomplish this ; and
that then some corn-crop should be grown for several
years in succession without manure, so as, as far as possible, to obliterate the unevenness of condition arising
from previous irregularity in manuring and cropping.

It will doubtless excite surprise when I say that,
notwithstanding the importance of the subject, and the
high authority on which the prosecution of scientifically conducted field-experiments has been advocated,
the conduct of such experiments has never been an
important part of the work of the Agricultural Experimental Stations on the continent of Europe, and that
it is now almost excluded from their programme.

In 1880, Professor Maercker of Halle, one of the
leading Agricultural Chemists of Germany, stated that
belief in their value was greatly diminished, and that
by some they were declared to be of no value. It is
objected that the Chemists of the Agricultural Stations
have neither the means nor the technical knowledge necessary for carrying out such experiments

successfully; that neither the amount of land, nor the funds at their disposal, are such as to admit of any safe deductions for application in practical agriculture from the results ; and that purely physiological problems can be better investigated in the laboratory, or in the greenhouse. He remarks that, owing to the errors necessarily incident to field-experiments conducted by those not acquainted with practical agriculture, the confidence of the practical farmer in the results has been shaken. Indeed, owing to the difficulties and cost of such enquiries, if conducted in a truly scientific manner, so as to be applicable for the solution of questions of fundamental and general interest, Professor Maercker concludes that the only field-experiments which it is practicable to carry out in Germany, are such as should be conducted by the practical farmer himself, to test the applicability to practice, of results and conclusions otherwise arrived at; and that, to insure that even such experiments are not misleading, similar ones should be conducted on different descriptions of soil, and for several years in succession.

I have already quoted the opinion of Sir Humphrey Davy, that scientifically conducted field-experiments should be undertaken by proprietors of land, who by their education are fitted to form enlightened plans, and by their fortunes are able to carry them into execution ; and when I tell you that the Rothamsted field-experiments, independently of all the laboratory investigations connected with them, cost considerably more, and those which have been undertaken by the Duke of Bedford at Woburn for the last seven years, on behalf of the Royal Agricultural Society of England, and which are under the direction of Dr. Voelcker, not much less, than £1000 annually, you will not be surprised that such field-experiments are not more general.

Prior to the appearance of Liebig's first work, in 1840, Mr., now Sir John Bennet Lawes, commenced experiments with different manuring substances, first with plants in pots, and afterwards in the field, at Rothamsted, into the hereditary possession of which he had entered on his majority in 1834. The results so obtained on a small scale in 1837, 1838, and 1839, were such as to lead to more extensive trials in the field in 1840, 1841, and subsequently.

In 1843, more systematic field-experiments were commenced; and a barn, which had already been applied to laboratory purposes, became almost exclusively devoted to agricultural investigations. These, which are still in progress, have been conducted entirely at the cost of Sir John Lawes, who has further set apart a sum of £100,000, and certain areas of land, for their continuance after his death.

In June 1843, I became associated with Mr. Lawes in the conduct of these investigations; and as it is doubtless my connexion with them (which still continues), to which my election to this Chair is mainly to be attributed, it will not be out of place to give, on this occasion, a brief outline of the scope and plan of the work which has been accomplished, during the more than forty years of its continuance. That I should do so seems the more desirable, since I interpret my appointment as indicating a feeling on the part of the Electors, that the results acquired in this long period of investigation of the scientific principles of agriculture, many of which are known by their publication, must have provided important material for illustration, in the lectures which it will be my duty to deliver. In fact, whilst I shall not neglect important results and conclusions established by others, the plan which I have in view will nevertheless involve much reliance on the data acquired at Rothamsted.

I may premise that since July 1855, when a new laboratory, built by public subscription of agriculturists, was presented to Mr. Lawes, the old barn laboratory has been abandoned. At the present time, the Rothamsted staff consists of two, and sometimes three chemists; several general assistants in different departments; occasionally a botanical assistant; three, and sometimes four, computors and record keepers; also a laboratory man, and several boys.

The general scope and plan of the field-experiments has been—to grow some of the most important crops of rotation, each separately, year after year, for many years in succession on the same land, without manure, with farmyard manure, and with a great variety of chemical manures; the same description of manure being, as a rule, applied year after year on the same plot. Experiments, with different manures on the mixed herbage of permanent grass-land, on the effects of fallow, and on an actual course of rotation, without manure, and with different manures, have likewise been made.

Field-experiments have thus been conducted for the periods, and over the areas, indicated in the table on the following page.

Samples of all the experimental crops are brought to the laboratory. Weighed portions of each are partially dried and preserved for future reference, or analysis. Duplicate weighed portions of each are dried at 100°C, the dry matter determined, and then burnt to ash. The quantities of ash are determined

Rothamsted Field Experiments.

CROPS.	DURATION.	AREA.	PLOTS.
	Years.	*Acres.*	
Wheat (various manures) . . .	41	13	37
Wheat, alternated with Fallow .	33	1	2
Wheat (varieties)	15	4–8	about 20
Barley (various manures) . . .	33	$4\frac{1}{4}$	29
Oats (various manures)	10 [1]	$\frac{3}{4}$	6
Beans (various manures) . . .	32 [2]	$1\frac{1}{4}$	10
Beans (various manures) . . .	27 [3]	1	5
Beans, alternated with Wheat . .	28 [4]	1	10
Clover (various manures) . . .	30 [5]	3	18
Various Leguminous Plants . .	7	3	17
Turnips (various manures) . . .	23 [6]	8	40
Sugar Beet (various manures) .	5	8	41
Mangel-Wurzel (various manures) .	9	8	41
Total Root Crops . .	42		
Potatoes (various manures) . . .	9	2	10
Rotation (various manures) . .	37	$2\frac{1}{2}$	12
Permanent Grass (various manures)	29	7	22

and recorded; the ashes themselves being preserved for reference, or analysis.

In a large proportion of the samples the total nitrogen is determined; and in some the amount existing as albuminoids, amides, and nitric acid.

In selected cases, illustrating the influence of season, manures, exhaustion, &c., complete ash-analyses have been made, numbering in all more than 700.

Also in selected cases, illustrating the influence of season and manuring, quantities of the experi-

---

[1] Including 1 year Fallow.
[2] Including 1 year Wheat, and 5 years Fallow.
[3] Including 4 years Fallow.
[4] Including 2 years Fallow.
[5] Clover, 12 times sown, 8 yielding crops, but 4 of these very small, 1 year Wheat, 5 years Barley, 12 years Fallow.
[6] Including Barley without manure 3 years (11th, 12th, and 13th seasons).

mentally grown wheat-grain have been sent to the mill, and the proportion and composition of the different mill-products has been determined.

In the sugar-beet, mangel-wurzel, and potatoes, the sugar in the juice has in many cases been determined by polariscope, and frequently by copper also.

In the case of the experiments on the mixed herbage of permanent grass-land, besides the samples taken for the determination of the chemical composition (dry matter, ash, nitrogen, woody fibre, fatty matter, and composition of ash), carefully averaged samples have frequently been taken for the determination of the botanical composition. In this way, on four occasions, at intervals of five years —viz., in 1862, 1867, 1872, and 1877—a sample of the produce of each plot was taken, and submitted to careful botanical separation; and the percentage, by weight, of each species in the mixed herbage determined. Partial separations, in the case of samples from selected plots (frequently of both first and second crops), have also been made in other years.

Samples of the soils of most of the experimental plots have been taken from time to time, generally to the depth of 9, 18, and 27 inches; sometimes to twice, and sometimes even to four times this depth. In this way more than 1000 samples have been taken, submitted to partial mechanical separation, and portions of the fine soil (that is, excluding stones) have been carefully prepared and preserved for analysis. In a large proportion of the samples the loss on drying at different temperatures, and at ignition, has been determined. In most, the nitrogen determinable by burning with soda-lime has been estimated. In many the carbon, and in some the nitrogen as nitric acid, and the chlorine, have been determined. Some experiments have also been made on the comparative absorptive capacity (for water and ammonia) of different soils and subsoils. The systematic investigation of the amount, and the condition, of the nitrogen, and of some of the more important mineral constituents, of the soils of the different plots, and from different depths, is now in progress, or contemplated.

Almost from the commencement of the experiments the rainfall has been measured—for more than thirty years in a gauge of one-thousandth of an acre area, as well as in an ordinary small funnel-gauge of 5 inches diameter. An 8-inch 'Board of Trade' copper-gauge is also now in use, commencing January 1, 1881. From time to time, the nitrogen as ammonia (and sometimes as nitric acid), has been determined in the rain waters. The chlorine has also been determined in many samples.

Three 'drain gauges,' also of one-thousandth of an acre each, for the determination of the quantity and composition of the water percolating, respectively through 20 inches, 40 inches, and 60 inches depth of soil (with its subsoil in natural state of consolidation), have also been constructed. Each of the differently manured plots of the permanent experimental wheat-field having a separate pipe-drain, the drainage-waters have been, and are frequently, collected and analysed.

The nitrogen existing as nitric acid, sometimes that in other forms, and also some other constituents, are, and for some time past have been, determined periodically, in both the rain and the various drainage waters.

For several years in succession, experiments were

made to determine the amount of water given off by plants during their growth. In this way various plants, including representatives of the gramineous, the leguminous, and other families, have been experimented upon. Similar experiments have also been made with various evergreen and deciduous trees.

Having regard to the difference in the character and amount of the constituents assimilated by plants of different botanical relationships, under equal external conditions, or by the same description of plants, under varying conditions, observations have been made on the character and range of the roots of different plants, and on their relative development of stem, leaf, &c. In the case of various crops, but more especially with wheat and beans, samples have been taken at different stages of growth, and the composition determined, in more or less detail, sometimes of the entire plant, and sometimes of the separated parts. In a few cases, the amounts of dry matter, ash, nitrogen, &c., in the above-ground growth of a given area, at different stages of development, have been determined. The amounts of stubble of different crops have also occasionally been estimated.

Experiments were made for several years in succession to determine whether plants assimilated free or uncombined nitrogen, and also various collateral points. Plants of the gramineous, the leguminous, and of other families, were operated upon. The late Dr. Pugh took a prominent part in this enquiry.

Obviously, an investigation of the scientific principles of Agriculture would be incomplete if it were not extended to the question of the feeding of the animals of the farm. Accordingly, experiments with such animals were commenced early in 1847, and have been continued at intervals up to the present time.

The following points have been investigated :—

1. The amount of food, and of its several constituents, consumed in relation to a given live-weight of animal, within a given time.

2. The amount of food, and of its several constituents, consumed to produce a given amount of increase in live-weight.

3. The proportion, and relative development, of the different organs or parts of different animals.

4. The proximate and ultimate composition of the animals, in different conditions as to age and fatness, and the probable composition of their increase in live-weight during the fattening process.

5. The composition of the solid and liquid excreta (the manure), in relation to that of the food consumed.

6. The loss or expenditure of constituents by respiration and the cutaneous exhalations—that is, in the mere sustenance of the living meat—and —manure-making machine.

Several hundred animals—oxen, sheep, and pigs—have been submitted to experiment.

The amount, and the relative development, of the different organs and parts, were determined in two calves, two heifers, fourteen bullocks, one lamb, 249 sheep, and 59 pigs.

The percentages of water, mineral matter, fat, and nitrogenous substance, were determined in certain separated parts, and in the entire bodies, of ten animals—namely, one calf, two oxen, one lamb, four sheep, and two pigs. Complete analyses of the ashes, respectively, of the entire carcases, of the mixed internal and other 'offal' parts, and of the entire bodies, of each of these ten animals have also been made.

From the data provided, as just described, as to the chemical composition of the different descriptions of animal, in different conditions as to age and fatness, the composition of the increase whilst fattening, and the relation of the constituents stored up in increase to those consumed in food, have been estimated.

To ascertain the composition of the manure in relation to that of the food consumed, oxen, sheep, and pigs, have been experimented upon.

The loss or expenditure of constituents, by respiration and the cutaneous exhalations, has not been determined directly, but only by difference; that is, by calculation, founded on the amounts of dry matter, ash, nitrogen, &c., in the food, and in the (increase) fæces, and urine.

Independently of the points of enquiry here enumerated, the results obtained have supplied data for the consideration of the following questions:—

1. The characteristic demands of the animal body, for nitrogenous or non-nitrogenous constituents of food, in the exercise of muscular power.

2. The sources in the food of the fat produced in the animal body.

3. The comparative characters of animal and vegetable food in human dietaries.

Supplementary investigations have also been made; for example—on the application of town-sewage to different crops, including experiments on the feeding qualities of the produce grown; the amount of increase yielded by oxen, and the amount and composition of the milk yielded by cows, being determined.

The chemistry of the malting process, the loss of food constituents during its progress, and the comparative feeding value of barley and malt, have also been investigated.

Many of the results of the investigations above enumerated have already been published, but a large proportion as yet remains unpublished.

As already intimated, I propose to rely largely on the data supplied in the more than forty years of investigation at Rothamsted, in the field, the feeding-shed, and the laboratory, in the lectures which it will be my duty to give in this University, in elucidation of the scientific principles of agriculture.

In the first place, however, to give logical sequence to what will follow, I shall devote a few lectures to the question of the sources of the constituents of the plants and animals which are the products of the art of agriculture. Obviously, the soil, and the atmosphere, are these sources. The subject of the origin, the general characters, and the composition, of soils, is a very wide one; and many lectures might be devoted to its discussion. But it is foreign to my plan so to discuss it; as to do so would seriously trench upon the time which should be devoted to the consideration of other branches of our enquiry.

I shall next consider the effects of manures, exhaustion, and variations of season, on the amounts of produce, and on the composition of the produce, of different crops, the effects of fallow, and the benefits arising from a rotation of crops.

Subsequently, I propose to take up the subject of the feeding of animals, for the production of meat, milk, and manure, and for the exercise of force—that is, for their labour. Also, if opportunity should occur, to treat of the question of the application of town-sewage to the land.

Finally, I should observe that, throughout the illustrations which I shall bring before you, it will be my endeavour to keep in view the bearing of the data with which I shall have to deal, on the important question of compensation for unexhausted improvements, so much in discussion at the present time.

The first experiments on crops to which I shall direct attention, will be those on the cereals; for, although bad seasons, and foreign competition, have of late tended to lessen the relative importance of these products to the British farmer, the experimental evidence obtained, both in the field and in the laboratory, is more complete in regard to them than to any other crops; and the discussion of the results, will afford the opportunity of considering important questions of wider interest to agriculture, than those exclusively relating to the production of the cereal crops themselves.

Printed at the University Press, Oxford
By HORACE HART, *Printer to the University*

www.ingramcontent.com/pod-product-compliance
Lightning Source LLC
Chambersburg PA
CBHW022024190326
41519CB00010B/1586